Final Countdown:

A World on the Brink

A. S. Aurelius

Contents

Introduction:

In the vastness of our universe, where mysteries abound and wonders never cease, humanity finds itself confronted with a horrifying reality. Imagine a world where the sky darkens, not with the shadows of passing clouds, but with the imminent approach of a colossal asteroid, a celestial harbinger of destruction. The news spreads like wildfire, piercing the hearts of every person on Earth with a chilling realization - we have a mere three years left before our world, as we know it, is irrevocably obliterated.

This book takes you on a journey into a future that no one wants to imagine, a future where the heavens themselves conspire against us. It is a story that explores the depths of human emotion, the resilience of the human spirit, and the fragile nature of our existence. Step into a world gripped by fear, anxiety, and sorrow, as the realization of an impending doom washes over us all.

We will venture into the heart of governments and witness their desperate attempts to establish protocols and strategies in the face of certain annihilation. The unfolding crisis will unravel the very fabric of society, as fear gives birth to chaos, lawlessness, and the breakdown of order. We will witness the struggle for survival, as scarcity grips the world and communities grapple with the question of how to sustain themselves in the final days.

As the countdown to our ultimate fate continues, we will delve into the complexities of human nature. Some will find solace in faith, seeking meaning in the midst of despair, while others will question the purpose of their daily routines and embrace the fleeting moments of joy with loved ones. We will witness the impact of the asteroid, as it unleashes cataclysmic forces upon our fragile planet, forever altering the face of the Earth.

But even in the darkest of times, a glimmer of hope remains. As the dust settles and the echoes of devastation fade, the surviving remnants of humanity will grapple with their place in the universe, pondering the lessons learned and embarking on a quest for knowledge that could safeguard future civilizations. We will explore the indomitable human spirit, as it rises from the ashes and sets its sights on a brighter future, fueled by the determination to preserve life against all odds.

While the events depicted in this book are born from the realm of imagination, they serve as a mirror reflecting our deepest fears, our capacity for resilience, and our quest for meaning in the face of impending doom. Welcome to a world where the final countdown begins, and humanity stands united in the face of its own mortality.

Chapter One

The Dreadful Discovery

In the vast expanse of space, where distant galaxies twinkle like celestial jewels, astronomers peer into the cosmos, their eyes fixed upon the endless depths. Among them, a dedicated group of scientists and researchers diligently scan the heavens, seeking answers to the enigmas that reside beyond our earthly domain.

It is within this diligent pursuit of knowledge that they make a discovery that sends shockwaves through the scientific community and reverberates across the globe. In the dark abyss of space, nestled within the reaches of our solar system, lies a celestial menace of unparalleled magnitude.

With eyes fixed upon the cosmos, these diligent observers witness the gradual emergence of an object, slowly making its way towards Earth. Through meticulous calculations and careful observation, they determine its true nature - an asteroid, a cosmic behemoth measuring a staggering 940 kilometers in diameter. The implications of this revelation send a shiver down their spines.

News of this monumental find spreads like wildfire, carried on the waves of advanced communication technologies. From the depths of research laboratories to the corridors of power, the information cascades, touching the lives of people from all walks of life. In an instant, the world is thrust into a state of collective shock, disbelief, and abject terror.

People turn their gaze skyward, their eyes straining to comprehend the magnitude of the impending doom hurtling toward them. As news outlets broadcast the dire announcement, the once-familiar constellations take on an ominous presence, casting a specter of impending death across the night sky. The realization sets in - humanity has a mere three years, a mere flicker of time, before the apocalyptic collision unfolds.

Fear grips the hearts of men, women, and children as the weight of the revelation bears down upon them. Waves of anxiety wash over the globe, leaving a trail of anguish and despair in their wake. Uncertainty permeates every corner of society, seeping into the very fabric of our collective consciousness.

Questions abound, whispered in hushed tones: Can this cataclysmic event be averted? Is there a way to divert or neutralize the approaching menace? The collective gaze turns to the institutions tasked with safeguarding humanity, foremost among them NASA's Jet Propulsion Laboratory, its "Asteroid Watch" bearing a solemn responsibility to detect and track these cosmic wanderers.

Within the corridors of power, governments scramble to assess the situation, seeking guidance from the experts, and formulating plans to protect their citizens. Emergency task forces assemble, crisis management strategies take shape, and contingency protocols are laid out in an effort to instill a semblance of order in the face of overwhelming chaos.

The weight of this knowledge hangs heavily upon the common person. Fear and anxiety permeate the air, as the stark reality sinks in - there is no escape, no place to hide from the impending devastation. The future, once a tapestry of dreams and aspirations, becomes a canvas painted with the hues of sorrow and finality.

As the world grapples with this newfound certainty, humanity finds itself united by a shared sense of vulnerability. Strangers become companions in grief, reaching out to one another for solace and support. The indomitable human spirit, resilient in the face of adversity, seeks solace in the embrace of loved ones, and yearns for a glimmer of hope in the encroaching darkness.

The clock is ticking, the celestial clockwork drawing ever closer to the final hour. The dreadful discovery echoes through the corridors of our existence, forcing us to confront the fragility of our world and ponder the depths of our own mortality.

Chapter Two

A World Unraveled

The dreadful discovery reverberates through the world, shattering the illusion of safety and stability. As the impending doom looms, society finds itself on the brink of chaos, its foundations trembling under the weight of uncertainty.

Governments, thrust into uncharted territory, scramble to establish protocols and navigate the treacherous path ahead. Emergency task forces work tirelessly, driven by urgency, to assess the situation and mitigate the impact of the approaching asteroid. In the United States, leaders address the nation, balancing transparency with the need to prevent mass panic, as the world watches and hangs onto every word.

Fear, an ever-present companion, spreads like wildfire, infecting hearts and minds. The realization of impending doom sends individuals on an emotional rollercoaster, questioning their mortality and the meaning of their actions. Some succumb to despair, overwhelmed by the weight of their impending demise, while others seek solace in the embrace of loved ones or in the comfort of their spiritual beliefs.

Society, once a tapestry of stability, begins to unravel at its seams. The strain of collective anxiety tests the pillars of law and order, as reports of looting, vandalism, and civil unrest seep through the cracks. Desperation erodes moral boundaries, blurring the lines between right and wrong in the face of imminent destruction.

Scarce resources become battlegrounds as panic buying grips communities. Grocery stores transform into scenes of frenzied chaos, stripped bare of essential supplies. Those without means or preparedness face the harsh reality of scarcity, struggling to secure even the most basic necessities.

Amidst the unraveling, however, flickers of resilience emerge. Communities unite, transcending divisions, and extending helping hands to those in need. Acts of kindness and compassion punctuate the darkness, reminding us of the strength and beauty that can blossom in the face of adversity.

The social contract that binds society reveals its fragility. Institutions once taken for granted falter under the strain, grappling with the delicate balance of maintaining order while ensuring the survival of their citizens.

The world teeters on the precipice. Uncertainty, fear, and dwindling trust permeate the air. Yet, amidst the chaos, a glimmer of resilience persists, reminding us of the indomitable spirit that resides within each of us. It is in these trying times that the true character of humanity is tested, as we navigate a landscape forever altered by the impending cataclysm.

Chapter Three

The Fractured Society

In the wake of the dreadful discovery and the unraveling of the world, society finds itself fractured, its once-cohesive fabric torn asunder. As the countdown to the asteroid's impact continues, the strains of fear, desperation, and uncertainty deepen the fault lines within communities.

The breakdown of law and order becomes an unfortunate reality, as the cracks in societal cohesion widen. Reports of looting, violence, and anarchy become all too common, as desperation and survival instincts override the remnants of moral compasses. People struggle to distinguish between right and wrong, and the thin veneer of civility that once held society together begins to crumble.

Scarcity reigns supreme as essential resources dwindle. Grocery stores resemble war zones, their shelves stripped bare by panicked individuals desperate to secure provisions for the uncertain future. The vulnerable, lacking means or preparedness, face the harshest consequences, struggling to find sustenance amidst the chaos.

The disintegration of societal norms amplifies the struggles faced by the marginalized and disadvantaged. Inequities and disparities that were already present become magnified, as resources become increasingly scarce and opportunities for survival diminish. The chasm between the haves and the have-nots widens, further deepening the fractures within society.

Communities, once united by shared goals and values, become divided along lines of self-interest and survival. Trust erodes, replaced by suspicion and skepticism. Neighbors, once familiar faces, become potential threats as individuals prioritize their own well-being above all else.

Institutions tasked with maintaining order and stability face monumental challenges. Governments struggle to strike a delicate balance between enforcing law and ensuring the survival of their citizens. Emergency measures are put in place, attempting to restore a semblance of control amidst the chaos. Yet, the task is daunting, as the cracks within society deepen and the forces of disorder gain momentum.

The fractured society stands as a somber reflection of the human condition under immense pressure. It exposes the inherent vulnerabilities and flaws within us,

As the fractures within society deepen, the consequences become increasingly dire. The breakdown of law and order creates a breeding ground for criminal activity. Gangs and opportunistic individuals exploit the chaos, further destabilizing communities and exacerbating the sense of fear and vulnerability. Theft, violence, and acts of desperation become commonplace, as the struggle for survival intensifies.

With the disintegration of societal norms, the concept of justice becomes distorted. Vigilante groups form, seeking to mete out their own form of retribution and protection. The lines between right and wrong blur, as individuals grapple with the harsh reality of their impending doom and the erosion of the systems that once governed their lives.

Food shortages worsen, leading to desperation and famine in many regions. Basic necessities become luxuries as scarcity reigns supreme. Those who can afford exorbitant prices or possess the means to secure resources by any means necessary find themselves in a position of privilege, while the most vulnerable bear the brunt of the suffering.

The fractured society also takes a toll on mental health. The constant state of fear and uncertainty weighs heavily on individuals, causing anxiety, depression, and despair to proliferate. The loss of stability and the impending threat of annihilation create a sense of hopelessness that hangs like a dark cloud over the collective psyche.

As societal structures continue to crumble, communication becomes increasingly fragmented. Traditional media outlets struggle to maintain credibility and provide accurate information amid the chaos. Rumors, misinformation, and conspiracy theories thrive in the void left by the breakdown of trusted sources. Uncertainty breeds confusion, further fracturing the already fragile social landscape.

Amidst the bleakness, some pockets of resilience persist. Communities band together, forming support networks and mutual aid systems to navigate the harsh realities of their new world. Acts of kindness and compassion become beacons of hope, serving as a reminder of the inherent goodness that can still be found in humanity's darkest hours.

Yet, the challenges that society faces are monumental. The fractures that have emerged threaten to dismantle the very foundations upon which civilization was built. It, becomes clear that the fractures within society are not easily mended, and the consequences of this unraveling will have far-reaching implications for the future of humanity.

Chapter Four

The Struggle for Survival

As the fractured society continues to grapple with the impending doom, a new chapter unfolds—one of relentless struggle for survival. We witness the lengths to which individuals and communities go in their desperate quest to outlast the impending cataclysm.

With resources dwindling and scarcity becoming the norm, the struggle for survival takes center stage. Every morsel of food, every drop of water, becomes precious currency in a world teetering on the edge of collapse. Individuals scavenge, barter, and fight for their share, driven by the primal instinct to preserve their own lives and those of their loved ones.

Communities that once thrived on cooperation now find themselves locked in fierce competition. Bonds of trust fracture further as survival becomes an individualistic pursuit. Neighbor turns against neighbor, as the lines between ally and adversary blur. In this struggle, the vulnerable bear the heaviest burdens, their pleas for assistance often drowned out by the cacophony of desperation.

The harsh realities of this new world force individuals to make unimaginable choices. Morality becomes a luxury, as people are confronted with decisions that test the limits of their humanity. Sacrifices are made, both small and profound, in the quest for continued existence. The line between right and wrong blurs further, as the survival instinct takes precedence over ethical considerations.

The breakdown of societal systems has far-reaching consequences. Healthcare and medical services become strained, unable to cope with the magnitude of the impending disaster. The sick and injured face unimaginable challenges, as the infrastructure that once supported their well-being crumbles. Those dependent on medication or specialized care find themselves in a precarious position, facing the harsh reality that their survival may be compromised.

The struggle for survival also gives birth to new forms of ingenuity and adaptation. Innovation becomes a necessity, as individuals and communities seek unconventional solutions to overcome the challenges they face. Makeshift shelters, alternative food sources, and survival strategies become the norm. The human capacity for resilience and creativity shines through, offering a flicker of hope in the face of overwhelming odds.

Yet, the struggle for survival exacts a heavy toll on the human spirit. Mental and emotional fortitude are tested as individuals grapple with the constant specter of death and uncertainty. Grief and loss permeate the air, as the magnitude of the impending disaster weighs heavily on hearts and minds. The struggle becomes not only physical, but also a battle for the preservation of hope and resilience.

The struggle for survival reveals the depths of human resilience and the lengths to which individuals will go in the face of imminent destruction. It showcases both the darkest aspects of human nature and the flickering flames of compassion and solidarity. In this fight for continued existence, the boundaries of what it means to be human are tested, forcing individuals to confront their own fragility and redefine their understanding of what truly matters.

Chapter Five

A World on Hold

As the struggle for survival persists, a sense of stagnation settles upon the world. In this chapter, we delve into the paradoxical nature of a world on hold—where time marches on, yet progress and normalcy seem suspended in the face of impending doom.

With the threat of the cataclysm looming ever closer, a collective sense of anticipation and dread pervades the air. People go about their daily routines, but with a heightened awareness of the limited time remaining. The rhythm of life becomes tinged with a bittersweet urgency, as individuals grapple with the paradox of continuing their existence while awaiting their eventual demise.

In this suspended reality, the pursuit of long-term goals and aspirations becomes fraught with uncertainty. Dreams of a brighter future, once driving forces behind human endeavor, lose their luster in the face of imminent destruction. Ambitions and plans, once pursued with fervor, now seem futile against the backdrop of the impending cataclysm. The world becomes suspended in a state of collective apprehension, as individuals grapple with the fleeting nature of their existence.

Governments, faced with unprecedented challenges, navigate uncharted territory. Protocols and plans are put in place to manage the impending crisis, yet the sheer magnitude of the threat renders many efforts seemingly futile. Resources are allocated towards emergency response and preparedness, but the overarching sense of doom casts a shadow over even the most well-intentioned endeavors. The machinery of governance, once a pillar of stability, finds itself strained under the weight of an uncertain future.

As the world remains on hold, the question of faith and spirituality takes center stage. Many seek solace in religious or philosophical beliefs, finding comfort and purpose in the face of the inexplicable. Places of worship become sanctuaries of hope, where individuals turn to prayer, meditation, or reflection to make sense of their impending fate. Others grapple with existential crises, questioning the existence of a benevolent higher power in the face of such impending devastation.

Amidst the suspension of normalcy, a spectrum of human emotions unfolds. Grief, sadness, and despair mingle with moments of unexpected joy and connection. Relationships are forged, mended, or fractured as individuals navigate the complexities of their own emotional landscapes. Love becomes both a source of solace and torment, as the impermanence of human existence casts a poignant shadow over even the most tender of connections.

The world on hold also reveals glimpses of resilience and the pursuit of beauty amidst chaos. Artists, musicians, and creators seek solace in their craft, channeling their emotions into works of art that capture the essence of the human experience. Beauty emerges as a beacon of hope, reminding individuals of the fleeting moments of grace that can be found amidst the turmoil.

Time takes on a dual nature. Days pass, seasons change, and life carries on, yet the weight of impending doom hangs heavy in the collective consciousness. People find solace in the mundane, cherishing simple pleasures and moments of respite. The world becomes a delicate balance of the temporal and the eternal, the fleeting and the everlasting.

We witness the profound paradox of a world on hold. Time continues its relentless march forward, yet progress and normalcy remain suspended. In this liminal space, individuals grapple with the existential and emotional toll of their impending fate. It is within this suspended reality that the complexities of the human experience are laid bare, forcing us to confront the fragility of our existence and search for meaning amidst the chaos.

In the suspension of normalcy, societal structures and institutions undergo a profound transformation. Economic systems teeter on the edge of collapse as uncertainty and fear grip the markets. Business operations become fragmented, as companies struggle to find purpose and sustainability in the face of impending doom. The pursuit of profit and growth loses its meaning, replaced by a collective focus on immediate needs and survival.

Education, once a cornerstone of societal progress, finds itself in disarray. Schools and universities grapple with the dilemma of preparing students for a future that may never come to fruition. Curriculum and pedagogy adapt, emphasizing practical skills and knowledge that can aid in survival. Intellectual pursuits take a backseat to practicality, as the world shifts its focus to immediate concerns.

In this suspended reality, the fabric of social relationships is both strained and strengthened. Some find solace in the comfort of familiar faces, drawing strength from the bonds of family and community. Others seek solitude, retreating into introspection and self-reflection. Loneliness and isolation become ever-present companions, as individuals grapple with their own mortality and the fragility of human connection.

The arts, too, undergo a metamorphosis in this world on hold. Creative expression becomes a form of catharsis, allowing individuals to process their emotions and make sense of their experiences. Writers pen poignant stories of hope and despair, musicians compose melodies that evoke both joy and sorrow, and visual artists capture the essence of the human condition in their brushstrokes.

As the world waits in limbo, the concept of time itself undergoes a transformation. Days blur into weeks, months into years, and the linearity of time loses its grip on human consciousness. The future becomes a hazy abstraction, and the present moment takes on a heightened significance. People live in the here and now, savoring each breath, each heartbeat, knowing that their time is finite.

Time moves forward, and yet progress and normalcy remain suspended in the face of impending doom. In this suspended reality, individuals grapple with their mortality, reevaluate their priorities, and seek solace in the simple pleasures of life. It is within this suspended space that the true essence of humanity is laid bare, as individuals navigate the emotional, social, and existential complexities of their existence.

Chapter Six

The Impact Unleashed

The has come, the cataclysmic event that humanity has been bracing for—the moment when the colossal asteroid finally makes its impact upon the Earth. The world holds its breath as the long-awaited reckoning arrives, unleashing devastation on an unimaginable scale.

The impact is a spectacle of destruction, a collision between celestial forces and the fragile planet we call home. The sheer magnitude of the event sends shockwaves reverberating across continents, as seismic waves ripple through the Earth's crust. The ground trembles violently, structures crumble, and landscapes are reshaped in an instant. The deafening roar of the impact fills the air, drowning out the cries of humanity.

The force unleashed by the colossal asteroid tears through the atmosphere, creating a cataclysmic explosion. The resulting shockwave expands outward, obliterating everything in its path. The energy released in this apocalyptic event is unparalleled, as the impact generates a blast wave capable of leveling cities, reducing them to rubble and dust.

In the aftermath, a thick cloud of dust and debris blankets the sky, blocking out the sun's rays. Darkness descends upon the world, shrouding it in an eerie twilight. Temperatures plummet as sunlight struggles to penetrate the dense veil above. The once-vibrant ecosystems, already ravaged by the impending doom, are pushed to their limits, as the delicate balance of nature is further disrupted.

Humanity, battered and bruised, emerges from the wreckage to face a new reality. The landscape is scarred, a testament to the power of the impact. Once-thriving cities now lie in ruins, their towering structures reduced to twisted metal and broken concrete. The remnants of civilization bear witness to the indomitable spirit of humanity, as pockets of survivors band together to navigate the desolate wasteland that was once their home.

The impact leaves behind a myriad of challenges that must be confronted. Scarce resources become even scarcer, as the infrastructure that supported the needs of a global population lies in ruins. The search for food, water, and shelter becomes a daily struggle, pushing individuals to their limits of endurance. The struggle for survival intensifies, as humanity fights to rebuild amidst the ashes.

In the aftermath of the cataclysmic impact, humanity grapples with the harrowing reality of its new existence. The once bustling cities now lay in ruins, reduced to a haunting silence.

The immediate aftermath is marked by chaos and confusion. People emerge from their hiding places, disoriented and shell-shocked, trying to comprehend the scale of devastation around them. The cries of anguish and despair fill the air, intermingled with the sound of desperate searches for loved ones lost in the aftermath.

The survivors face the grim reality of limited resources. The destruction of infrastructure has severed supply chains, leaving them grappling with shortages of food, clean water, and medical supplies. They scavenge amidst the debris, salvaging what they can to sustain themselves and their communities. Re-sourcefulness becomes a vital skill, as they adapt to their new reality and learn to make do with the remnants of the past.

Amidst the struggle for survival, a new social order begins to emerge. Leadership, once anchored in governmental structures, undergoes a transformation. Charismatic individuals step forward, offering guidance and rallying the survivors with their resilience and vision. Makeshift communities form, with each member contributing their skills and expertise to ensure the collective well-being.

But not all is unity and cooperation. Desperation breeds conflict as survivors compete for scarce resources. Fractures emerge within communities, testing the bonds that hold them together. The struggle for power and control becomes an ever-present undercurrent, threatening to undermine the fragile progress made amidst the ruins.

Amidst the challenges, the remnants of technology become a lifeline. The survivors salvage remnants of communication systems, enabling them to establish rudimentary networks. These networks become a vital tool for information exchange, allowing communities to share knowledge, coordinate efforts, and provide support to one another.

As days turn into weeks and weeks into months, hope flickers amidst the darkness. The survivors' resilience and adaptability begin to bear fruit as they witness small signs of progress. Makeshift shelters are constructed, rudimentary farming practices are implemented, and rudimentary healthcare systems are established. These efforts, though modest, provide a glimmer of optimism in the face of overwhelming odds.

Nature, too, shows signs of rejuvenation amidst the ruins. Amidst the barren landscapes, sprouts of green emerge, a symbol of nature's resilience and ability to reclaim what was lost. Wildlife, driven to the brink of extinction, finds refuge in pockets untouched by the cataclysm. The survivors draw inspiration from these glimpses of nature's tenacity, finding solace in the beauty and resilience that endures even in the face of unimaginable destruction.

The impact unleashed reshapes the world, both physically and emotionally. The struggle for survival becomes a relentless battle, where humanity's resolve is tested at every turn. Yet, amidst the challenges and hardships, the indomitable human spirit prevails. Communities form, innovation thrives, and hope persists, as humanity strives to rebuild and carve out a new existence amidst the shattered remnants of the past.

Chapter Seven

Aftermath and Reflection

The remnants of humanity now face the task of rebuilding amidst the ruins, while grappling with the profound existential questions that arise in the wake of such devastation.

As the dust settles and the initial shock subsides, the survivors begin to assess the scope of the damage left behind by the impact. The once-thriving cities are reduced to ghostly shells, with remnants of shattered infrastructure serving as haunting reminders of a bygone era. The landscapes bear the scars of the cataclysm, forever transformed by the immense forces unleashed upon them.

Amidst the destruction, communities band together, drawing strength from their shared experiences. They pool their resources, knowledge, and skills, working tirelessly to establish a semblance of normalcy. The rebuilding efforts are marked by resilience, as survivors tap into their collective ingenuity to construct shelters, cultivate food, and develop systems to meet their basic needs.

In the process of rebuilding, reflection becomes an essential part of the survivors' journey. The magnitude of the impact forces them to confront existential questions about the nature of life, the fragility of existence, and the inherent impermanence of all things. These questions weigh heavily on their hearts as they grapple with grief, loss, and the search for meaning in a world forever changed.

The survivors find solace in coming together to share their stories and experiences. They create spaces for dialogue and reflection, where they can process their emotions and find support in one another. These moments of connection and shared vulnerability become a source of healing, helping to rebuild not only the physical structures but also the shattered spirits.

In the aftermath, the survivors also grapple with the question of responsibility. They confront the harsh reality that the cataclysmic impact was not merely a random act of nature but a result of human negligence and shortsightedness. They reflect on the choices that led them to this point, questioning the systems and values that prioritize individual gain over collective well-being. This introspection prompts a collective commitment to creating a more sustainable and compassionate world moving forward.

The survivors become custodians of knowledge, preserving the lessons learned from the devastation. They establish educational initiatives to ensure that future generations understand the consequences of their actions and the importance of stewardship. Scientific research becomes a cornerstone of their efforts, as they seek to better understand the world and prevent similar catastrophes in the future.

As the survivors rebuild and reflect, they find hope amidst the ruins. They witness the resilience of the human spirit and the capacity for compassion and cooperation that emerged in their darkest hours. They find strength in the bonds forged through adversity and vow to honor the memory of those lost by creating a world that cherishes life and protects the planet.

The survivors stand at the precipice of a new era. They carry with them the scars of the past, but also the wisdom gained from their experiences. With determination and a newfound sense of purpose, they embark on a journey of healing, renewal, and the collective pursuit of a brighter future.

In the process of reconstruction, the survivors prioritize sustainability and environmental stewardship. They recognize that the impact served as a stark reminder of the fragile interdependence between humanity and the planet. With this realization, they embrace sustainable practices, seeking to restore balance and heal the wounded Earth. Renewable energy sources become the backbone of their efforts, as they harness the power of sun, wind, and water to meet their energy needs.

Communities organize themselves into decentralized networks, fostering resilience and adaptability. They establish local governance systems that empower individuals to actively participate in decision-making processes. These decentralized structures foster a sense of ownership and accountability, ensuring that the diverse needs and perspectives of the survivors are represented.

Education becomes a cornerstone of the rebuilding efforts. Recognizing the importance of knowledge and critical thinking, the survivors establish schools and learning centers. They prioritize scientific literacy, environmental education, and the exploration of ethics and values. Through education, they strive to cultivate a generation that is equipped to navigate the complexities of a post-impact world and make informed choices for the betterment of humanity.

In the midst of reconstruction, the survivors engage in deep introspection and personal growth. They contemplate the existential questions that the impact thrust upon them. With newfound humility, they embrace the impermanence of life and seek meaning in the present moment. They find solace in the interconnectedness of all beings, cherishing each other and the natural world with a profound sense of gratitude.

Art and creativity flourish in the wake of the impact, serving as a means of expression, healing, and catharsis. Survivors channel their emotions into various art forms, creating powerful works that reflect the depths of their experiences. Music, literature, painting, and performance become vital vehicles for story-telling and the preservation of their collective memory.

Throughout the journey of rebuilding and reflection, the survivors draw strength from their shared humanity. They recognize that compassion, empathy, and collaboration are essential for the collective well-being. They actively foster a culture of inclusivity, embracing diversity and celebrating the richness of human experiences. Through this unity, they forge a path toward a more equitable and just society.

The survivors stand on the threshold of a transformed world. They have weathered unimaginable challenges and emerged with a renewed sense of purpose and interconnectedness. With the scars of the impact etched upon their landscapes and hearts, they march forward with resilience, hope, and a deep appreciation for the preciousness of life.

The survivors recognize that the impact, though devastating, has sparked a collective awakening—a profound shift in consciousness that has the power to reshape their world for the better.

Chapter Eight

The Search for Solutions

In this chapter, we delve into the tireless efforts of the survivors to search for solutions in the wake of the cataclysmic impact. Faced with the daunting task of rebuilding their world, they embark on a journey of innovation, collaboration, and scientific exploration.

Recognizing the need for long-term survival, the survivors turn their attention to finding ways to prevent future catastrophes. They pool their collective knowledge and resources, establishing research institutions dedicated to studying celestial bodies and monitoring potential threats from space. Cutting-edge technology and advancements in space exploration become crucial tools in their pursuit of a safer future.

One key organization at the forefront of this effort is the reinvigorated NASA's Jet Propulsion Laboratory (JPL) and its Asteroid Watch program. Collaborating with international partners, JPL focuses on developing more sophisticated detection systems and refining asteroid tracking techniques. Through their tireless work, they strive to identify and characterize potential threats, providing early warnings and invaluable data for mitigation efforts.

The survivors recognize the need for a comprehensive planetary defense strategy. They invest in the development of advanced technologies capable of diverting or mitigating potential impactors. Concepts such as kinetic impactors, gravity tractors, and solar sails are explored, offering potential methods to alter the trajectory of hazardous asteroids. International collaborations in this endeavor become paramount, as nations come together to pool their expertise and resources.

While scientific advancements are crucial, the survivors also acknowledge the importance of public awareness and education. They launch global campaigns to inform and engage the general population about the potential threats posed by near-Earth objects. Citizen science initiatives flourish, with individuals contributing to asteroid detection, tracking, and characterization efforts. Through education and awareness, the survivors hope to empower communities worldwide to actively participate in safeguarding their future.

The search for solutions extends beyond Earth itself. The survivors embrace the concept of space colonization as a means of ensuring the long-term survival of humanity. Ambitious projects are launched, focused on establishing self-sustaining habitats on other celestial bodies such as the Moon or Mars. These off-world settlements serve as insurance policies, providing alternative homes for future generations should Earth ever face another devastating impact.

However, the survivors also grapple with ethical dilemmas in their pursuit of solutions. They confront questions of resource allocation, access, and the potential consequences of altering celestial bodies. Debates arise surrounding the responsibility humanity holds as caretakers of the cosmos and the potential impact of our actions on the delicate balance of the universe.

As the survivors strive to find solutions, the importance of international cooperation becomes increasingly evident. Global treaties are established to govern planetary defense efforts, ensuring equitable access to resources and knowledge. Collaborative platforms are formed, enabling scientists, engineers, and policymakers from around the world to share information, exchange ideas, and coordinate mitigation strategies.

The survivors stand on the precipice of a transformed future. Through their unwavering determination, they have made significant strides in their search for solutions. The specter of future celestial threats remains, but they face it with resilience, ingenuity, and a collective commitment to safeguarding humanity and the planet.

The survivors navigate the complexities of space exploration, scientific advancements, and ethical considerations. They embark on a quest for solutions, driven by the desire to protect future generations and ensure the long-term survival of humanity. In their united efforts, they seek to unlock the secrets of the cosmos and forge a path towards a safer and more resilient future.

Fueled by shared purpose, survivors continue to explore the vast expanse of space in search of answers and strategies to protect Earth. Their tireless efforts and unwavering commitment serve as a testament to the indomitable spirit of humanity. In their relentless pursuit, they strive not only to prevent future cataclysms but also to forge a legacy of resilience and hope that transcends the boundaries of time and space.

Epilogue: A Glimmer of Hope

We reflect on the remarkable journey of the survivors, the progress they have made, and the glimmer of hope that shines through the darkness. Despite the immense challenges and devastating impact, humanity emerges stronger, united, and with a newfound appreciation for the preciousness of life.

In the aftermath of the cataclysmic event, the survivors continue their efforts to rebuild and heal. The scars left by the impact serve as a constant reminder of the fragility of their existence, urging them to forge a path toward a more sustainable and compassionate world.

The lessons learned from the near-destruction of Earth propel the survivors to embrace a harmonious coexistence with nature. They implement eco-friendly practices, restoring ecosystems, and replenishing the planet's resources. Sustainable agriculture, renewable energy, and conservation efforts become integral parts of their daily lives, ensuring a balanced and thriving environment.

The survivors understand the importance of unity and collaboration as they navigate the complexities of their new reality. Borders dissolve as nations come together, sharing knowledge, resources, and technologies for the betterment of all. Trust and cooperation become the foundation upon which a brighter future is built.

Through their collective efforts, the survivors establish resilient communities that prioritize inclusivity, equality, and justice. They dismantle the remnants of a fragmented society and foster a culture of empathy and compassion. No longer bound by divisions, they recognize the inherent value of every individual, nurturing a society that celebrates diversity and cherishes the inherent worth of all human beings.

Education becomes a beacon of hope for the survivors, empowering future generations with the knowledge and skills to navigate the challenges of a post-impact world. Schools and universities thrive, offering a curriculum that intertwines scientific understanding, ethical reasoning, and cultural appreciation. The pursuit of knowledge becomes a driving force, as they strive to unlock the mysteries of the universe and understand their place within it.

As the survivors gaze upon the night sky, they find solace in the vastness of the cosmos. They cultivate a renewed sense of wonder and awe, recognizing their interconnectedness with the universe. The impact serves as a catalyst for a profound shift in their perspective, fostering a deep reverence for life and a commitment to preserving the cosmic tapestry of existence.

Amidst the healing and rebuilding, a glimmer of hope emerges. It is the realization that humanity possesses an innate resilience, an unwavering spirit that refuses to be extinguished. They understand that while the threat of celestial impacts remains, they have the knowledge, technology, and unity to face the challenges ahead.

In the final moments of this journey, the survivors embrace a future filled with hope, possibility, and a deep appreciation for the fragile beauty of life. They carry the lessons learned from their tumultuous past and embark on a collective endeavor to nurture a world that is resilient, sustainable, and filled with compassion.

As we conclude this book, we are reminded that the story of the survivors is not just one of tragedy and devastation, but also of resilience, unity, and the triumph of the human spirit. It is a testament to the capacity of humanity to overcome the darkest of challenges and find hope in the face of uncertainty.

Though the journey ahead may be fraught with obstacles, the survivors remain steadfast in their commitment to protect Earth and its inhabitants. They carry within them a glimmer of hope that serves as a guiding light, inspiring them to forge a future where humanity thrives in harmony with the cosmos, forever grateful for the gift of life.

Conclusion

In this journey through the impending doom of an asteroid impact, we have witnessed the tumultuous path humanity takes when faced with the inevitability of destruction. The vivid descriptions and emotional narratives have allowed us to delve into the depths of fear, resilience, and the unwavering human spirit.

Throughout the pages of this book, we have painted a picture of a world on the brink of collapse. We have felt the palpable anxiety and witnessed the unraveling of societies, as individuals grapple with their impending fate. The struggle for survival and the harrowing aftermath have laid bare the depths of human desperation and determination.

But amidst the chaos, there have been glimpses of hope. The survivors, in their darkest hours, have come together, united by a shared desire to preserve what remains of their world. Their unwavering spirit and ingenuity have allowed them to find solace in collective action, working tirelessly to rebuild and adapt.

The narrative has shown the power of collaboration and the strength of the human will. It has highlighted the importance of scientific knowledge and technological advancements in the face of existential threats. From the shattered remnants of society, new communities have emerged, built upon the principles of compassion, equality, and a deep reverence for life.

As the story reaches its conclusion, we are left with a profound appreciation for the fragility of our existence and the resilience of the human spirit. We are reminded that even in the face of unimaginable devastation, hope can arise. It is a reminder that, as stewards of our planet, we must cherish and protect the gift of life.

This book serves as a poignant call to action. It urges us to reevaluate our priorities, to nurture a sustainable coexistence with nature, and to foster a sense of global unity. Through education, collaboration, and forward-thinking, we can navigate the challenges of an uncertain future and steer humanity towards a more resilient and compassionate world.

In the face of the unknown, we must embrace hope, draw on our collective wisdom, and work together to safeguard our planet and the legacy we leave behind.

Acknowledgments

Thank you for taking the time to read my book. If you have found it enjoyable or beneficial, I would greatly appreciate it if you could spare a few moments to write a brief review on Amazon Kindle. Your support means a lot to me, and it truly makes a difference. I genuinely value your feedback as it enables me to grow and enhance my work in the future. Thank you.